Junior Mathematical Team Games

Enjoyable activities to enhance the junior curriculum

Vivien Lucas

tarquin

How to organise Mathematical Team Games

Photocopy a page of the chosen game for each team. Cut them out and then shuffle them a little. Teams can then deal them out in the normal way. Some teachers might prefer to photocopy them on to coloured paper and perhaps to laminate the sheets to give a more permanent and significant feel to the cards. I usually make seven or eight copies, each on a card of a different colour.

Such a method would mean that a collection of team games could be ready to use at very short notice indeed. It also means that you have a second game in reserve if the first one ends too soon.
These games have been tried on primary school pupils from 8 to 11 with very positive feedback. The secondary school version - *Mathematical Team Games* - is hugely popular with pupils and teachers.

Why are the Star cards numbered?

The numbers have no significance within the context of the team-game itself but are there simply to check that none of the cards are missing. In these games the final question will be found on one and possibly on two of the cards. As a positive decision in this collection, the question will never be found on card 1.

Since each card does contain some unique piece of information, a single missing card would mean that the problem cannot be solved. None of them are duds or dummies.

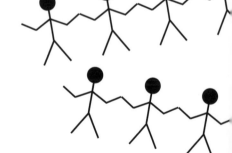

This game is based on lines of symmetry of shapes, for example a square has 4 lines of symmetry.

Reflection Story

What is the best size for the teams?

Although in theory it does not matter if the number of cards dealt to each player is different, since the information is to be shared anyway, in practice it is better that each team member makes the same contribution. Luckily the division properties of the duodecimal system are a great help here.

Teams of three or four are probably the most satisfactory in any case and these sizes should be your first choice. If you do have to have a team of five, then it is best to insist that that the two spare cards be placed in the centre of the table.

It is of course possible to divide the class into teams of two as they would get six cards each. However, a team of two does not really encourage the social interaction that this approach to mathematical teaching is intended to produce.

Ownership of the cards

Try to encourage everyone to retain ownership of their cards and the information that they contain. The aim is to encourage cooperation and the ownership of certain pieces of information gives a status to everyone present and helps to ensure that nothing is forgotten or missed. Everyone holds some part of the jigsaw and contributes to the success of the project. In fact the process is very like tackling an ordinary jigsaw puzzle without having the picture on the box. Initially it is not immediately clear even what the problem is or how to start to work towards what the answer might be.

Sorting out the muddle and bringing order out of chaos is an important part of the satisfaction that these team-games offer.

Drawing Lots

It is probably a good idea to draw lots publicly to determine who should be in each of the teams. These team-games also act as good icebreakers for new groups or for new entrants to an established class. The randomness of the selection process helps the process of making new friends and contacts.

Competition between teams

Games like these encourage healthy competition between the teams and add an element of time pressure to finding the solution. Small edible prizes usually go down well.

How much help to give

This will depend very much on the ability of the group but I usually only allow questions along the lines of 'Is this right so far?'

Solving the problems

An important part of the value and enjoyment of these team games lies in picking out and recognising the significant pieces of information on the star-cards. Then in arranging this information and in constructing a table, chart or diagram to get to the point where the mathematics can be done and the solution found. Finally, all the information can be used and all the loose ends tied up. It will often be the case, just as in so many real life situations 'Once you can formulate the question properly, the answer is obvious'.

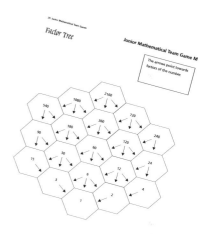

See the Tarquin website for more ideas about designing your own team games using the templates in this book: search for *Junior Mathematical Team Games* and follow the links below the description.

The Family Tree

This is the Answer sheet for the Family Tree Problem.

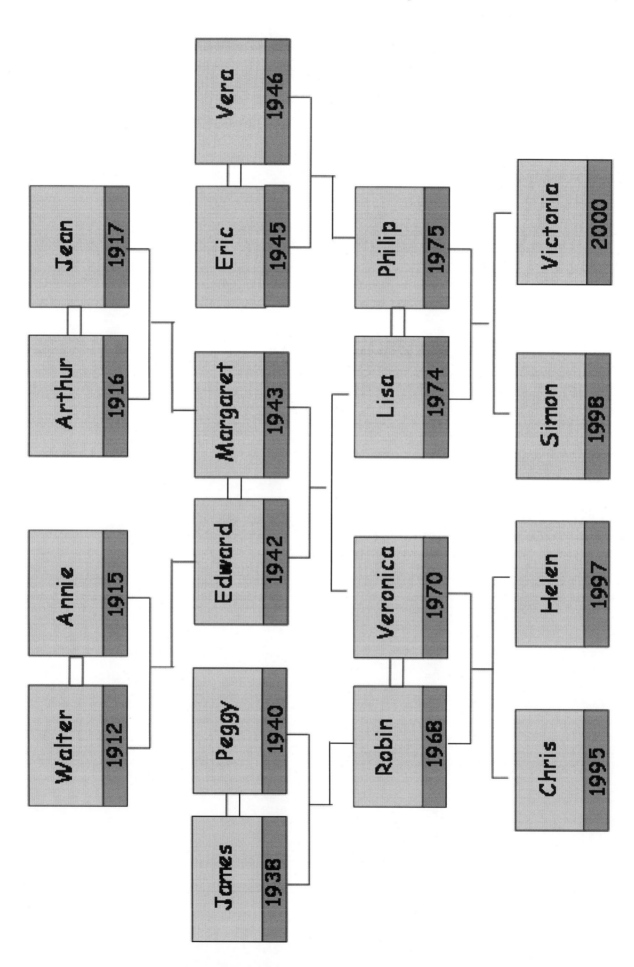

The Family Tree

Instructions

This activity uses words of comparison like younger, older, etc and involves addition and subtraction of four digit numbers (years of birth).

Topic:	Addition and
	Subtraction
Ages:	Any

Each team needs:

The set of game cards.

A family tree diagram with dates on.

You could provide name cards to place on the diagram made by photocopying and cutting up the answer sheet or they could just write the names onto the family tree.

What the pupils have to do

They have to read the information on the game cards and put the names in the correct places on the family tree diagram.

Follow up questions

1. If they are all still alive, how old would they be this year?
2. What is the relationship between…. E.G. Eric and Simon (grandfather and grandson).
3. How old was Edward when Chris was born?

Follow up Activities

Pupils could make their own family tree.

The Family Tree

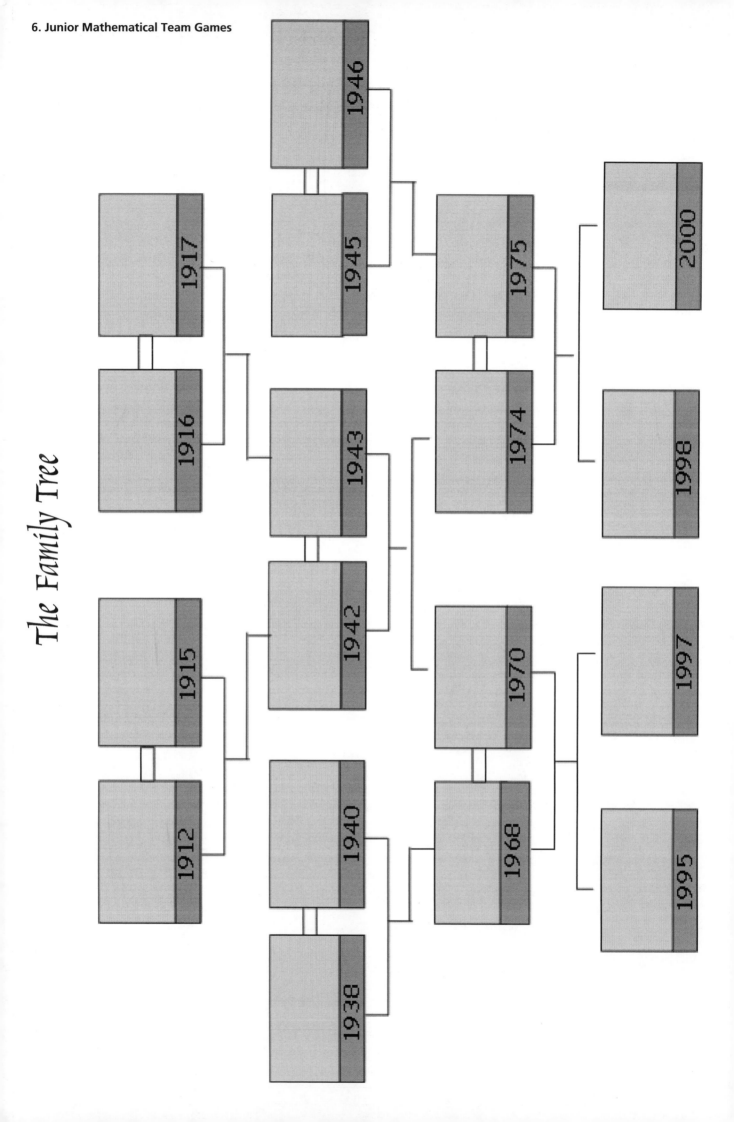

1 The family tree has 4 generations with the great grandparents at the top and the youngest generation at the bottom.

The Family Tree

2 Your task is to place all the family members in the correct places on the family tree.

The Family Tree

3 Victoria's mother is called Lisa and Lisa is the sister of Veronica and their parents are Edward and Margaret.

The Family Tree

4 Helen is one year older than Simon & 3 years older than Victoria. Chris is 21 years younger than his Auntie Lisa.

The Family Tree

5 Chris, Helen, Victoria and Simon are cousins and they decided to investigate their family tree.

The Family Tree

6 Chris and Helen are the children of Robin and Veronica. Helen was born when Robin was 31.

The Family Tree

7 Of the four cousins, Chris is the oldest and Victoria is the youngest. She was born in the year 2000.

The Family Tree

8 Margaret was the only child of Jean and Arthur and she is one year younger than her husband.

The Family Tree

9 Lisa is 4 years younger than her sister Veronica and 1 year older than her husband Philip.

The Family Tree

10 Robin's parents are James and Peggy. James is 60 years older than Simon. Peggy is 5 years older than Eric.

The Family Tree

11 The oldest family member is Walter, who was 88 when Victoria was born. He is married to Annie.

The Family Tree

12 Veronica is married to Robin who is 2 years older than her. Annie was 80 when Chris was born.

The Family Tree

13 Walter and Annie are Edward's parents and Walter was 30 when Edward was born.

The Family Tree

14 Victoria's mother Lisa was 26 when Victoria was born.

The Family Tree

15 Philip's parents are Eric and Vera. Eric is 30 years older than Philip and Vera is 3 years younger than Margaret.

The Family Tree

16 The cousins share one set of grandparents and two sets of great grandparents.

The Family Tree

The Magic Square

Instructions

This activity uses different types of numbers. Pupils need to know the meanings of odd, even, multiple, digit, prime, palindromic, perfect square and the concept that a magic square has all rows, columns and diagonals adding up to the same total.

Topic: Types of Numbers
Ages: 9–11

Each team needs:

The set of game cards.

Some blank grids to experiment with.

What the pupils have to do

They have to read the information on the game cards and put the numbers into the correct places on the magic square diagram.

Follow up questions

1. What is the sum of the 4 corners? (Answer 34)
2. What is the sum of the four centre squares? (34)
3. What other sets of four numbers add up to 34?

BLANK GRID

	1	2	3	4
A				
B				
C				
D				

ANSWERS

	1	2	3	4
A	1	8	10	15
B	12	13	3	6
C	7	2	16	9
D	14	11	5	4

The Magic Square

1

The multiples of 4 are in A2, B1, C3 and D4.

The Magic Square

2

The perfect squares are in A1, C3, C4 and D4.

The Magic Square

3

The single digit numbers are in A1, A2, B3, B4, C1, C2, C4, D3 and D4.

The Magic Square

4

A1 + B1 = B2

The Magic Square

5

There is a two digit palindromic number in D2.

The Magic Square

6

The even numbers are in A2, A3, B1, B4, C2, C3, D1 and D4.

The Magic Square

7

The numbers 1 to 16 appear once each and each row and column adds up to 34.

The Magic Square

8

The multiples of 3 are in A4, B1, B3, B4 and C4.

The Magic Square

9

The multiples of 5 are in A3, A4 and D3.

The Magic Square

10

The prime numbers are in B2, B3, C1, C2, D2 and D3. (Remember 1 is not prime).

The Magic Square

11

B3 + B4 = C4

The Magic Square

12

D2 − D3 = B4

Crack the Code

This puzzle is based on multiplication tables, but also solving codes which require pupils to know their alphabet. The tables used are 2, 3, 4 & 5. You could give pupils a copy of the alphabet and squared paper is also useful to line up the letters.

Topic: Times tables

Ages: 8–11

ANSWERS

Ben used the 2 times table (DAD BAKED A CAKE)

A	B	C	D	E	F	G	H	I	J	K	L	M
B	D	F	H	J	L	N	P	R	T	V	X	Z

Dean used the 3 times table (WE ALL LOVE MATHS)

A	B	C	D	E	F	G	H	I	J	K	L	M
C	F	I	L	O	R	U	X	A	D	G	J	M

N	O	P	Q	R	S	T	U	V	W	X	Y	Z
P	S	V	Y	B	E	H	K	N	Q	T	W	Z

Amy used the 4 times table (JACK HAD A BAD HEADACHE)

A	B	C	D	E	F	G	H	I	J	K	L	M
D	H	L	P	T	X	B	F	J	N	R	V	Z

Carol used the 5 times table (JILL HID A COOK BOOK)

A	B	C	D	E	F	G	H	I	J	K	L	M	N	O
E	J	O	T	Y	D	I	N	S	X	C	H	M	R	W

A group of four friends decided to make up coded messages for each other based on times tables.

1 Crack the Code

It is your task to work out which person chose which table (from 2 to 5) and then de-code the messages.

2 Crack the Code

In each message, the letters of the alphabet are given numbers from a = 1, b = 2, c = 3 up to z = 26.

3 Crack the Code

In the 2 times table code, each letter gets coded to its double, so A becomes B, B becomes D, up to M → Z.

4 Crack the Code

In the 3 times table code, each letter gets coded to 3 times itself, so A becomes C, B becomes F & so on.

5 Crack the Code

In the 4 times table code, each letter gets coded to 4 times itself, so A becomes D, B becomes H & so on.

6 Crack the Code

In the 5 times table code, each letter gets coded to 5 times itself, so A becomes E, B becomes J & so on.

7 Crack the Code

If you get a number bigger than 26, just repeat the alphabet again, so A = 27, B = 28 and so on.

8 Crack the Code

Amy's code is
NDLR FDP D HDP FTDPDLFT

9 Crack the Code

Ben's code is
HBH DBVJH B FBVJ

10 Crack the Code

Carol's code is
XSHH NST E OWWC JWWC

11 Crack the Code

Dean's code is
QO CJJ JSNO MCHXE.

12 Crack the Code

Maths Town

Instructions

This activity uses an alphabetical code: a = 26, b = 25 ...... z = 1. Pupils need to be able to do the basic four rules of number and know the meaning of squared. They have to do a sum and check the answer against their code which will give a letter which tells them where to put each of the Town's amenities on the map.

Topic: Codes &
Mathematical
Vocabulary
Ages: 9–11

Each team needs:

The set of game cards.

A copy of the Maths Town map with letters on.

Working out paper.

What the pupils have to do

They have to read the information on the game cards, do the sums, decode the answer and put the amenities into the correct places on the map.

After completing the map

Follow up questions

1. Describe the route from Integer Inn to Compass Café.
2. What does Integer mean? (Whole number)
3. What place is south of the Prime Police Station and also west of St. Cross Church? (High Hospital)
4. Discuss the properties of the shapes used e.g. hexagon.

Follow up Activities

1. Pupils could describe routes for other pupils to work out the destinations.
2. Pupils could design their own Maths Town using as many mathematical words as possible.

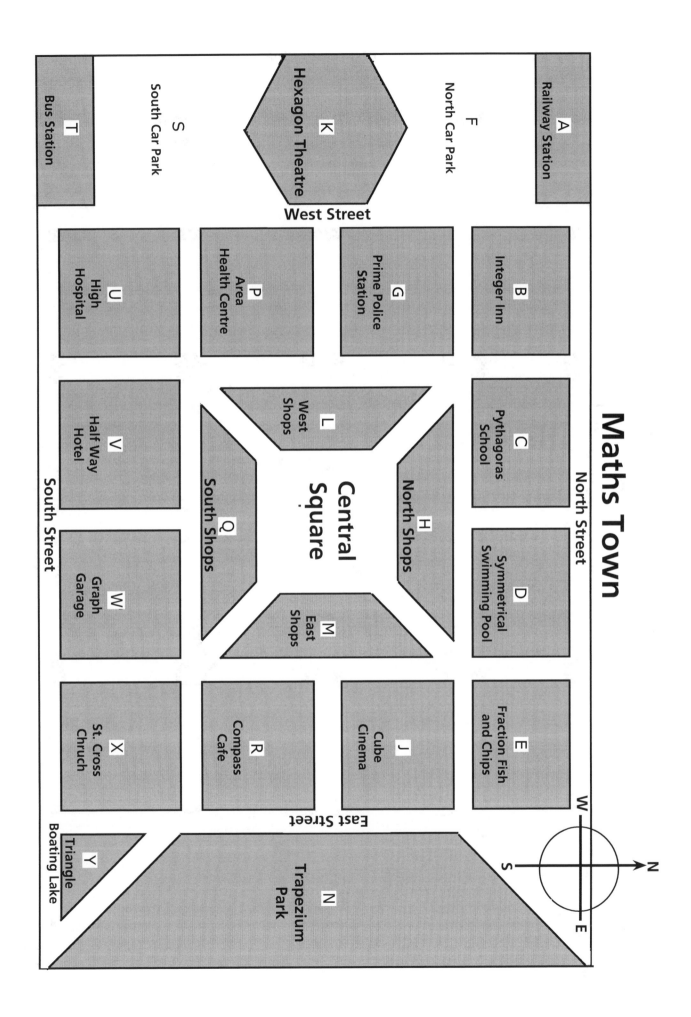

Maths Town

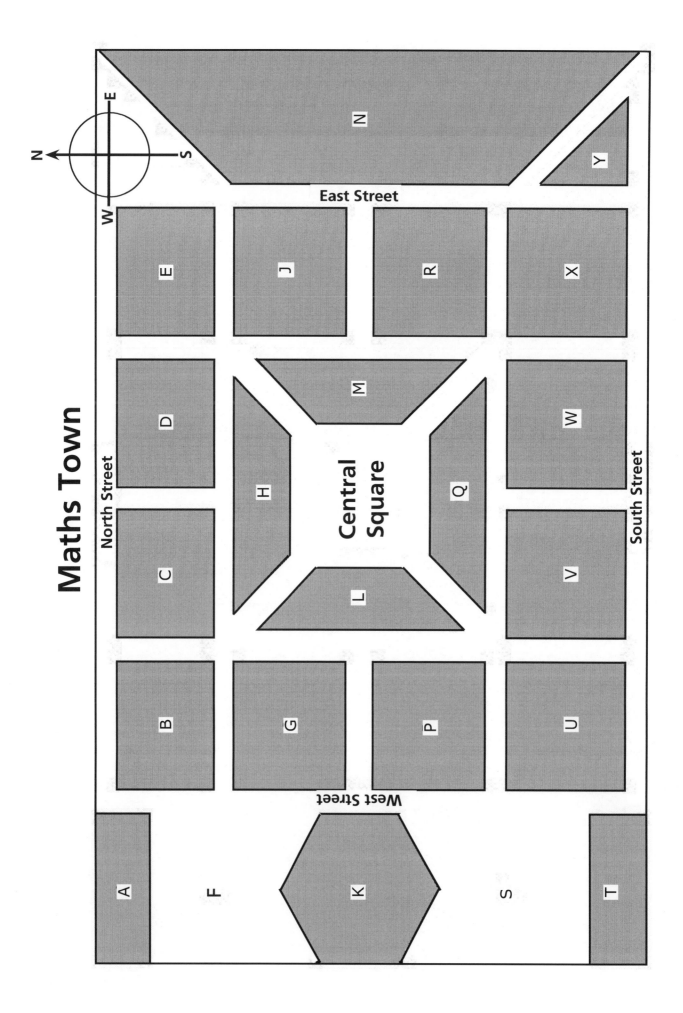

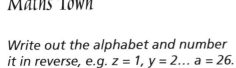 **Maths Town**

Write out the alphabet and number it in reverse, e.g. z = 1, y = 2... a = 26. This is the code you need to solve the problem.

 Maths Town

Each place in Maths Town has a letter showing its position. You have to work out where everything is from the clues.

 Maths Town

The railway station is at 2 x 13. Symmetrical Swimming Pool is at 36 – 13.

 Maths Town

Graph Garage is at 20 ÷ 5. The Area Health Centre is at 100 – 89.

 Maths Town

Prime Police Station is at 4 x 5. Pythagoras school is at 8 x 3. Integer Inn is at 5 squared.

 Maths Town

The Hexagon Theatre is at 4 squared. The north shops are at 55 – 36.

 Maths Town

The Half Way Hotel is at 45 ÷ 9. The south shops are at 80 ÷ 8.

 Maths Town

Trapezium Park is at 52 ÷ 4. Compass café is at 3 squared.

 Maths Town

The High Hospital is at 42 ÷ 7. The east shops are at 93 – 79.

 Maths Town

The north car park is at 7 x 3. St. Cross Church is at 36 ÷ 12.

 Maths Town

The south car park is at half of 16. The west shops are at 3 x 5. The Triangular Boating lake is at half of 4.

 Maths Town

The Cube Cinema is at 66 – 49. The Bus Station is at 63 ÷ 9. Fraction Fish & Chips is at 2 x 11.

Going Shopping

This a problem about money, in which the pupils have to work out how money is spent by each person in the story, how much they started with and how much they have left. They may need more than one copy of the answer grid, because the answers have to be arranged in the order of the amounts they spent.

Topic: Money

Ages: any

GOING SHOPPING ANSWERS

Name	Started with	Money left	Amount spent
Neal	£10.00	£0.50	£9.50
Oliver	£15.00	£6.00	£9.00
Carla	£8.00	nothing	£8.00
Hadia	£12.00	£5.20	£6.80
Angela	£6.50	£1.30	£5.20
Neleema	£7.20	£2.20	£5.00
Graham	£4.40	£0.80	£3.60
Eddie	£5.00	£2.25	£2.75

The message is NO CHANGE.

GOING SHOPPING ANSWER GRID

Name	Started with	Money left	Amount spent

1 *Going Shopping*

Angela's class went on a trip to a chocolate factory to find out how chocolate was made.

2 *Going Shopping*

While there, they were allowed to spend some money in the factory shop.

3 *Going Shopping*

You have to work out what they each spent and arrange them in order of amount spent to find a message.

4 *Going Shopping*

Oliver started off with the most money, he had a £5 note & 5 £2 coins. Carla was the only one to spend all her money.

5 *Going Shopping*

Hadia started off with a £10 note, 2 50p coins & 5 20p coins. Neleema bought a £5 box of chocolates for her Mum.

6 *Going Shopping*

Angela had a fifth of her money left after she bought sweets totalling £5.20. Neal went home with just 50p left.

7 *Going Shopping*

Graham had 1 each of 50p, 20p & 10p left after he bought a mug for his Dad. Eddie had saved his pocket money, of 50p a week for 10 weeks.

8 *Going Shopping*

Neal had £10 to spend on presents for his twin sisters. Eddie had just 3 coins left, a £2, a 20p coin and a 5p coin after he had shopped.

9 *Going Shopping*

Oliver just had to buy a teddy for £9. Hadia bought a £5 box of chocolates and 6 postcards for her collection at 30p each.

10 *Going Shopping*

Carla bought a toy bus, advertising the chocolate, which she paid for with 6 £1 coins, 3 50p coins, 2 20p coins and one 10p coin.

11 *Going Shopping*

Neleema had saved her pocket money for the trip, 80p a week for 9 weeks. Graham's mug cost him £3.60.

12 *Going Shopping*

Angela took £1.50 more than Eddie with her on the trip. The total amount of money spent that day was £49.85.

Arriving At School

The clues tell the pupils about how a group of pupils travel to school. It is all about adding or subtracting minutes from times of day to work out what time each pupil left home, arrived at school and how long they took. They may need more than one copy of the answer grid to work with, because the final answer has to be arranged in the order in which they arrive at school. Some pupils might need a clock face to help them work out the answers. There are opportunities for follow up work here, for example plotting graphs of journey times or modes of transport. The pupils could write stories about the people in the game. They could do a similar survey of the pupils in their own class.

Topic: Time
Ages: 8–11

ARRIVING AT SCHOOL ANSWER GRID

Name	Time they leave home	Time they arrive at school	Journey time in minutes	Method of transport

ANSWERS

Name	Time they leave home	Time they arrive at school	Journey time in minutes	Method of transport
Jane	8.14	8.20	6	Walk
Ursula	8.15	8.27	12	Car
Simon	8.20	8.30	10	Cycle
Tracey	8.25	8.33	8	Walk
Ian	8.15	8.39	24	Bus
Nigel	8.10	8.45	35	Car
Tony	8.30	8.50	20	Cycle
Isabella	8.51	8.55	4	Walk
Mike	8.34	8.58	24	Train & walk
Emma	8.25	9.05	40	Bus

The message is JUST IN TIME. Emma was late for school and Ian and Emma did not catch the same bus.

Junior Mathematical Team Game F

Arriving At School

Ursula and Ian both leave home at 8.15am. Nigel leaves home first, 41 minutes earlier than Isabella who leaves home last.

Arriving At School

Tracey and Emma both leave home at 8.25am. Nigel travels by car taking 35 minutes. Jane has a 6 minute walk to school.

Arriving At School

Simon leaves home at 8.20, just as Jane arrives at school. The bus stop is right outside school. Tony arrives at 8.50am.

Arriving At School

Ian and Emma both travel by bus. Mike's journey time is the same as Ian's. Isabella leaves home at 8.51am.

Arriving At School

Ian and Emma normally catch the same bus, but do they this time? Tracey has an 8 minute walk to school.

Arriving At School

Tony leaves home at 8.30, just as Simon arrives at school. Nigel arrives at 8.45am. Emma arrives 45 minutes after Jane.

Arriving At School

Emma's bus journey is 16 minutes longer than Ian's. Simon takes 10 minutes to cycle to school.

Arriving At School

Tony leaves home 4 minutes earlier than Mike. Jane is the first to arrive, 13 minutes before Tracey.

Arriving At School

Nigel's car journey is 5 minutes quicker than Emma's bus. Ian's bus journey takes twice as long as Ursula's car ride.

Arriving At School

Tony's cycle ride takes twice as long as Simon's. Mike has a 20 minute train ride and a 4 minute walk to get to school.

Arriving At School

Emma's bus journey takes 10 times as long as Isabella's walk. School starts at 9am. Find out if anyone arrives late.

Arriving At School

Arrange the names in the order in which they arrive and the first letter of each will spell out a message.

Old, Older, Oldest

This game is about working out the ages of all the people in the story and then working out the mean, median and the mode. The clues are mainly about people being either older or younger than other people. It is a good idea to suggest that they arrange their answers in descending order of age. This will help them to find the median.

Topic: Mean, median, mode

Ages: 10–11

ANSWERS

Name	Age
Steven	18
Maria	15
Jack	14
Haleema	12
Dennis (twin)	11
George (twin)	11
Brian	10
Fayruz	9
Paul	8
Ruth	7
Emma	6
Chris	5
Amy	4

Total of ages = 130 years

Mean = 10, Median = 10, Mode = 11.

Junior Mathematical Team Game G

Your task is to work out the ages of the 13 people and work out the mean, median and the mode.

1 Old, Older, Oldest

Amy is the youngest, Steven is the oldest and Brian is the seventh oldest.

2 Old, Older, Oldest

There is one set of male twins and there are 4 people older than the twins and 7 people younger.

3 Old, Older, Oldest

Emma is one year older than Chris, 1 year younger than Ruth and 2 years younger than Paul.

4 Old, Older, Oldest

Steven is twice as old as Fayruz and 3 times as old as Emma, who has 2 people younger than her.

5 Old, Older, Oldest

Ruth is half the age of Jack and Maria is 3 times the age of Chris, who is the second youngest.

6 Old, Older, Oldest

Paul is 8 years old, Amy is 4 and Maria, the second oldest, is 3 years younger than Steven.

7 Old, Older, Oldest

The ages of Chris, Dennis, George and Ruth are all prime numbers and their 4 ages add up to 34.

8 Old, Older, Oldest

Steven is 14 years older than Amy and all 13 ages add up to 130 years.

9 Old, Older, Oldest

Steven's age is equal to the ages of Emma and Haleema added together and they are all even numbers.

10 Old, Older, Oldest

Brian is 10 years old, Haleema is 12 and no-one is 13, 16 or 17.

11 Old, Older, Oldest

Fayruz, Emma, Haleema, Maria and Steven all have ages that are multiples of 3.

12 Old, Older, Oldest

The Tower of Hex

Instructions

This activity uses multiplication and also division with remainders. Calculators could be used but they don't give remainders so they are not really an advantage. Pupils need to know how many days in each month.

> **Topic:** Multiplication, division and calendar dates
>
> **Ages:** 9–11

Each team needs:

The set of game cards.

Working out paper.

What the pupils have to do

They have to read the information on the game cards, do the sums and work out the date by which the Tower of Hex was completed.

SOLUTION:

Number of bricks needed = 600 x 60 = 36000

At 6 bricks per hour, number of man hours = 6000

At 6 hours a day, number of man days = 1000

With 12 workers, number of days = 83.333 days (round to 84)

The date on the 84th working day will be Saturday 5th June.

Follow up questions

1. What words do you know that start with Hex?

2. What might have happened if humans had been made with 6 fingers on each hand and 6 toes on each foot?

Follow up Activities

1. A discussion on everything to do with the number 6, that it is even, a triangular number and a perfect number (6 = 1 + 2 + 3) the sum of the first 3 integers and the sum of its factors excluding itself.

2. Pupils could design their own Hexite. (Or Octogite etc.)

Junior Mathematical Team Game H

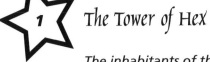 **1** The Tower of Hex

The inhabitants of the planet Hex are called Hexites.

 **2** The Tower of Hex

The inhabitants of the planet Hex decide to build a big tower.

 **3** The Tower of Hex

The workers work for 6 hours each day and they work a 6 day week, resting on Sundays.

 **4** The Tower of Hex

Twelve workers were employed to build the tower.

 **5** The Tower of Hex

The Hexites have 6 fingers on each hand and 6 toes on each foot.

 **6** The Tower of Hex

Your task is to work out the date they finished it.

 **7** The Tower of Hex

The Hexites started work on Monday 1st March.

 **8** The Tower of Hex

The Tower had 60 layers of bricks.

 **9** The Tower of Hex

Each layer of the tower required 600 bricks.

 **10** The Tower of Hex

Each worker was able to lay 6 bricks each hour.

 **11** The Tower of Hex

On the planet Hex, they have many hexagonal buildings.

 **12** The Tower of Hex

The area around the tower was paved with a tessellation of hexagonal slabs.

Reflection Story

This game is based on lines of reflection symmetry. There is story to complete by working out the missing words. Each letter of each missing word is represented by a shape and the pupils have to work out the number of lines of symmetry of each shape and then refer to the code table to work out the letters. There may need to be some discussion of line symmetry before they attempt this problem. Pupils need a copy of the code table and a copy of the story with gaps in.

Topic: Line symmetry

Ages: 10–11

Letters	A	R	E	F	L	E	C	T	I	O	N
Lines of Symmetry	0	1	2	3	4	2	5	6	8	16	>16

STORY ANSWER SHEET

1	and	2	3	friends.	They
both	enjoy	Maths.	Today	1	sat
4	5	6	7	and	2
sat	8	the	window.	They	had
9	10	how	9	11	5
6	12	They	were	13	9
14	the	shapes	15	16	helped.
1	asked	2	"How	many	12 s
17	symmetry	5	6	18	is
16	19	20	21 ?"	2	replied
"No, I	think	16	is	22	

STORY ANSWERS

Carol	and	Ellen	are	friends.	They
both	enjoy	Maths.	Today	Carol	sat
alone	in	a	corner	and	Ellen
sat	near	the	window.	They	had
to	learn	how	to	reflect	in
a	line.	They	were	free	to
trace	the	shapes	if	it	helped.
Carol	asked	Ellen	"How	many	lines
of	symmetry	in	a	circle,	is
it	nine	or	ten ?"	Ellen	replied
"No, I	think	it	is	infinite.	

Junior Mathematical Team Game I

This game is based on lines of symmetry of shapes, for example a square has 4 lines of symmetry.

1 *Reflection Story*

Your task is to complete the words in the story by solving the clues. Each word has a number (1–22).

2 *Reflection Story*

Each letter is represented by a picture. Count the lines of symmetry and look at the code.

3 *Reflection Story*

Word 1 is
Word 2 is

4 *Reflection Story*

Word 3 is
Word 4 is
Word 5 is

5 *Reflection Story*

Word 6 is
Word 7 is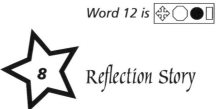

6 *Reflection Story*

Word 8 is
Word 9 is
Word 10 is

7 *Reflection Story*

Word 11 is
Word 12 is

8 *Reflection Story*

Word 13 is
Word 14 is

9 *Reflection Story*

Word 15 is
Word 16 is
Word 17 is

10 *Reflection Story*

Word 18 is
Word 19 is
Word 20 is

11 *Reflection Story*

Word 21 is
Word 22 is

12 *Reflection Story*

The Fraction Game

Instructions

The fraction game answer sheet needs copying onto paper, one per team. The sheets with the sections to colour need photocopying and cutting up into 20 strips, one set for each team. The fraction code sheet could be copied onto card and laminated.

Topic: Fractions
Ages: 9–11

Each Team needs:

A copy of the answer sheet to fill in.

20 separate fraction strips to colour in.

A copy of the fraction code sheet.

Colouring pencils.

What the pupils have to do:

Share out the fraction strips and colour in the fraction stated starting from the bottom of the strip where the numbers are. When all the strips have been coloured, they are compared with the fraction diagrams on the fraction code sheet and then the equivalent fractions and the letters are written on the fraction game answer sheet, which is handed to the teacher for checking. The message should read FRACTIONS ARE GREAT FUN.

FRACTION GAME ANSWERS

$\frac{2}{8}=\frac{1}{4}$	$\frac{6}{16}=\frac{3}{8}$	$\frac{5}{10}=\frac{1}{2}$	$\frac{2}{6}=\frac{1}{3}$	$\frac{14}{16}=\frac{7}{8}$	$\frac{2}{12}=\frac{1}{6}$	$\frac{2}{16}=\frac{1}{8}$	$\frac{10}{12}=\frac{5}{6}$	$\frac{10}{16}=\frac{5}{8}$		$\frac{4}{8}=\frac{1}{2}$	$\frac{9}{24}=\frac{3}{8}$	$\frac{4}{6}=\frac{2}{3}$
F	**R**	**A**	**C**	**T**	**I**	**O**	**N**	**S**		**A**	**R**	**E**

	$\frac{6}{8}=\frac{3}{4}$	$\frac{15}{40}=\frac{3}{8}$	$\frac{8}{12}=\frac{2}{3}$	$\frac{3}{6}=\frac{1}{2}$	$\frac{21}{24}=\frac{7}{8}$		$\frac{3}{12}=\frac{1}{4}$	$\frac{2}{10}=\frac{1}{5}$	$\frac{20}{24}=\frac{5}{6}$
	G	**R**	**E**	**A**	**T**		**F**	**U**	**N**

Follow up questions

Why is 2/8 the same as 1/4? Etc.

What must I add to 3/8 to make a whole?

What did you notice about (1) 2/8 and (18) 3/12?

Extra Activity

Using the fraction code sheet, the teacher or the pupils can make up more words for the pupils to work out, for example:

$\frac{2}{6}$	$\frac{3}{24}$	$\frac{10}{12}$	$\frac{8}{12}$
C	O	N	E

FRACTION GAME ANSWER SHEET

$\frac{2}{8} =$	$\frac{6}{16} =$	$\frac{5}{10} =$	$\frac{2}{6} =$	$\frac{14}{16} =$	$\frac{2}{12} =$	$\frac{2}{16} =$	$\frac{10}{12} =$	$\frac{10}{16} =$		$\frac{4}{8} =$	$\frac{9}{24} =$	$\frac{4}{6} =$

$\frac{6}{8} =$	$\frac{15}{40} =$	$\frac{8}{12} =$	$\frac{3}{6} =$	$\frac{21}{24} =$		$\frac{3}{12} =$	$\frac{2}{10} =$	$\frac{20}{24} =$

The Fraction Code

Letter	Fraction
A	$\frac{1}{2}$
C	$\frac{1}{3}$
E	$\frac{2}{3}$
F	$\frac{1}{4}$
G	$\frac{3}{4}$
I	$\frac{1}{6}$
N	$\frac{5}{6}$
O	$\frac{1}{8}$
R	$\frac{3}{8}$
S	$\frac{5}{8}$
T	$\frac{7}{8}$
U	$\frac{1}{5}$

Junior Mathematical Team Game K

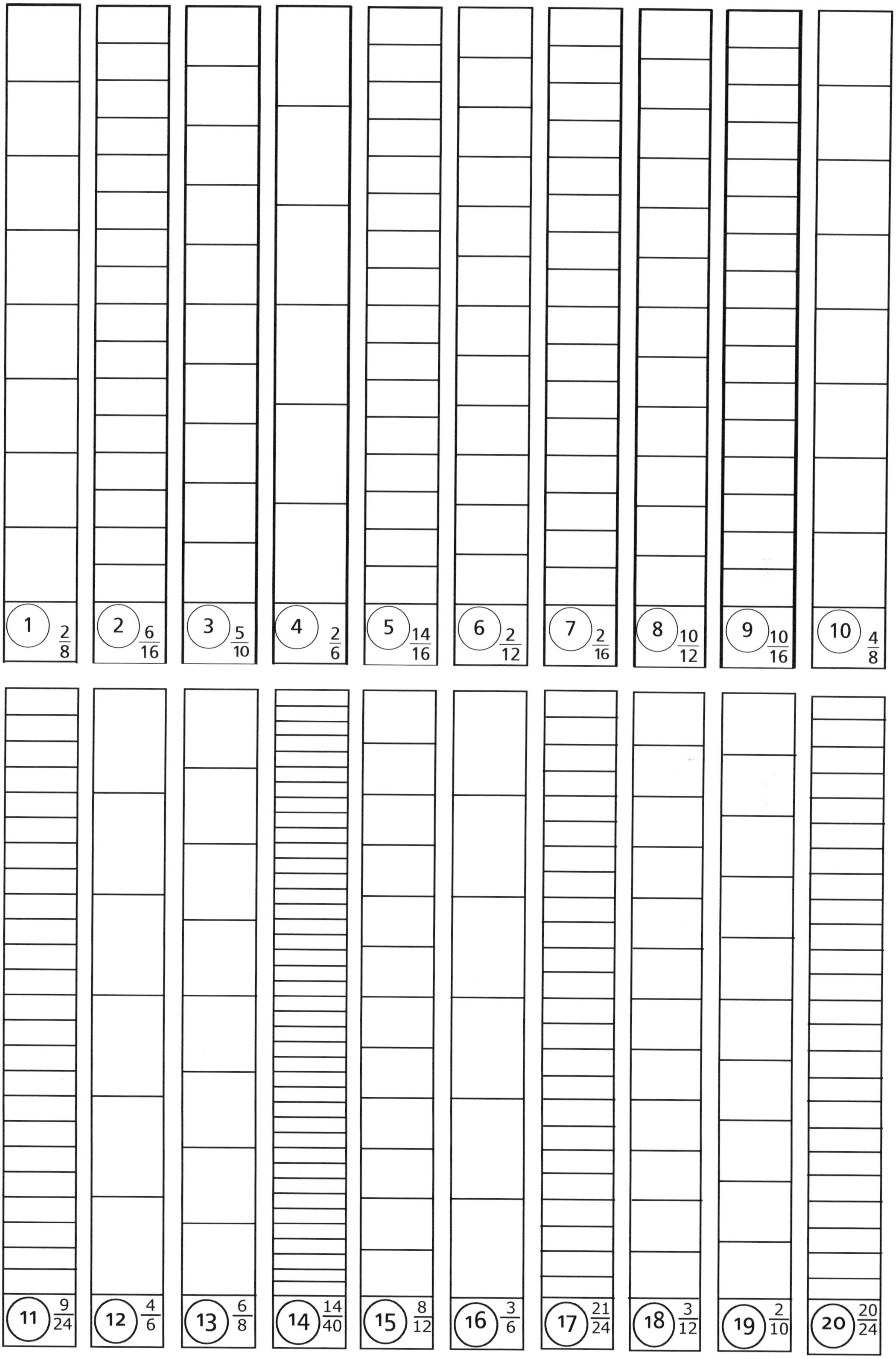

① 1 $\frac{2}{8}$

② 2 $\frac{6}{16}$

③ 3 $\frac{5}{10}$

④ 4 $\frac{2}{6}$

⑤ 5 $\frac{14}{16}$

⑥ 6 $\frac{2}{12}$

⑦ 7 $\frac{2}{16}$

⑧ 8 $\frac{10}{12}$

⑨ 9 $\frac{10}{16}$

⑩ 10 $\frac{4}{8}$

⑪ 11 $\frac{9}{24}$

⑫ 12 $\frac{4}{6}$

⑬ 13 $\frac{6}{8}$

⑭ 14 $\frac{14}{40}$

⑮ 15 $\frac{8}{12}$

⑯ 16 $\frac{3}{6}$

⑰ 17 $\frac{21}{24}$

⑱ 18 $\frac{3}{12}$

⑲ 19 $\frac{2}{10}$

⑳ 20 $\frac{20}{24}$

The Hexagon Jigsaws

These activities are easy to organise. First photocopy the jigsaws onto card, enough for one puzzle for each team of two to four pupils, and cut up the pieces. You could set the photocopier on an enlarged setting to create larger jigsaw pieces. You could also laminate the pieces for longer life.

All the teams have to do is to complete the puzzles in the fastest time possible. Some of the puzzles contain a message which is there for easy checking or the pupils could write it down and bring it to the teacher as soon as they have finished.

A blank hexagon grid is available for you to make up your own jigsaw puzzles. The matching numbers can either go in pairs at the centre of each side or in threes at the corners of the hexagons. The puzzles could also be given to individual pupils to work on alone if they have finished the set work.

There are four puzzles included here:

1. Addition and subtraction

2. Multiplication and division

3. Make 100 (pairs of numbers that add up to 100)

4. Factors (In each hexagon the arrows that point from each number point towards factors of that number).

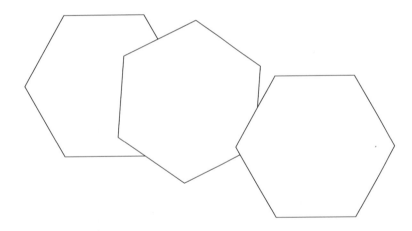

Junior Mathematical Team Game L

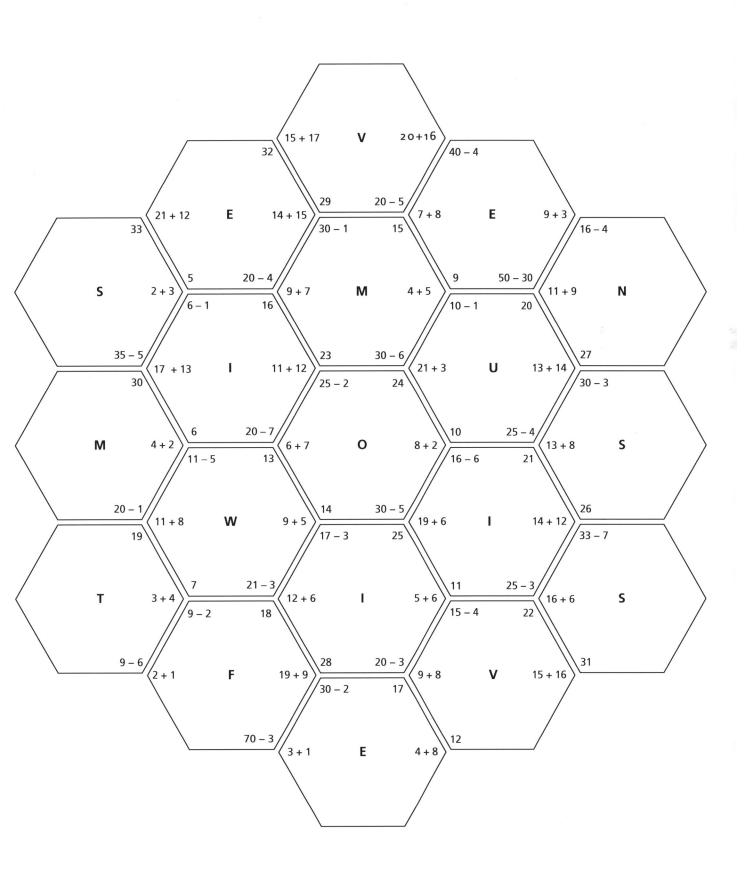

Junior Mathematical Team Game L

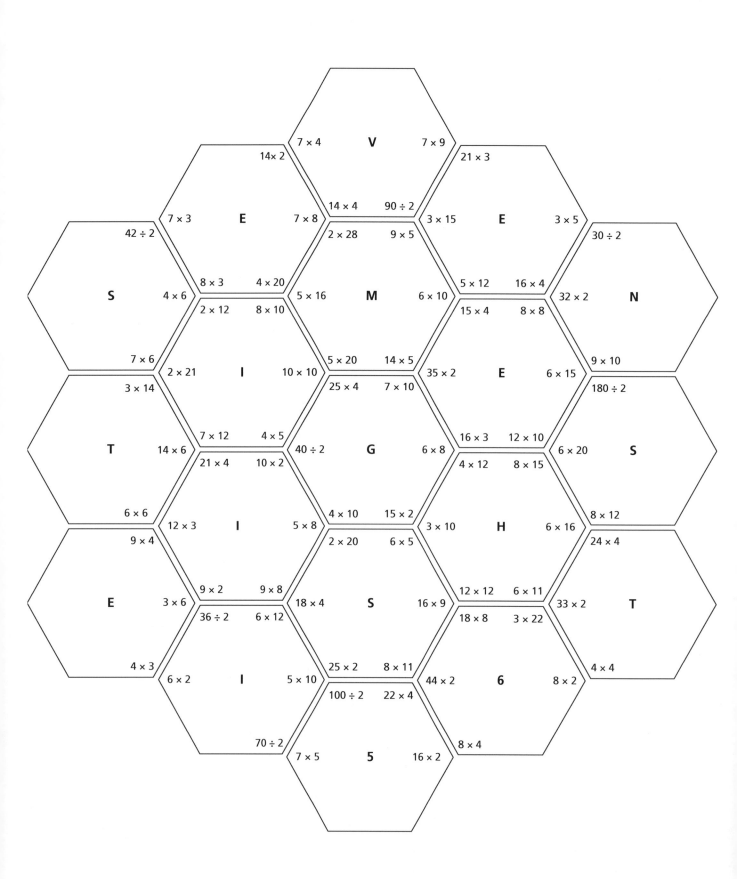

Make 100

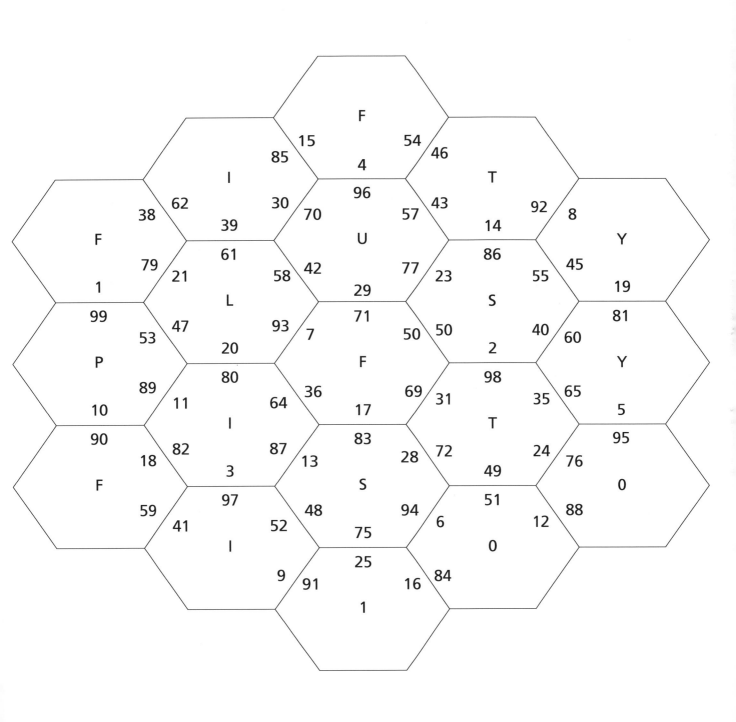

Factor Tree

The arrows point towards factors of the number.

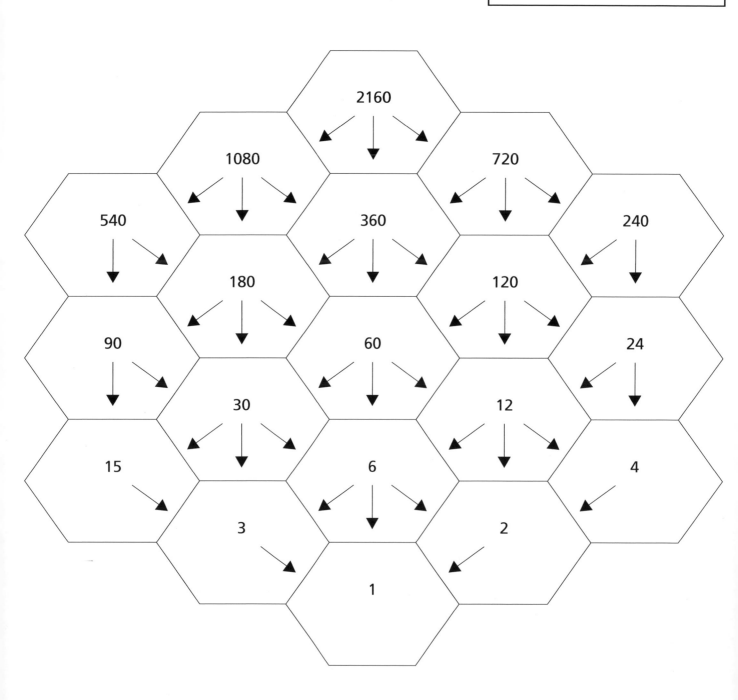

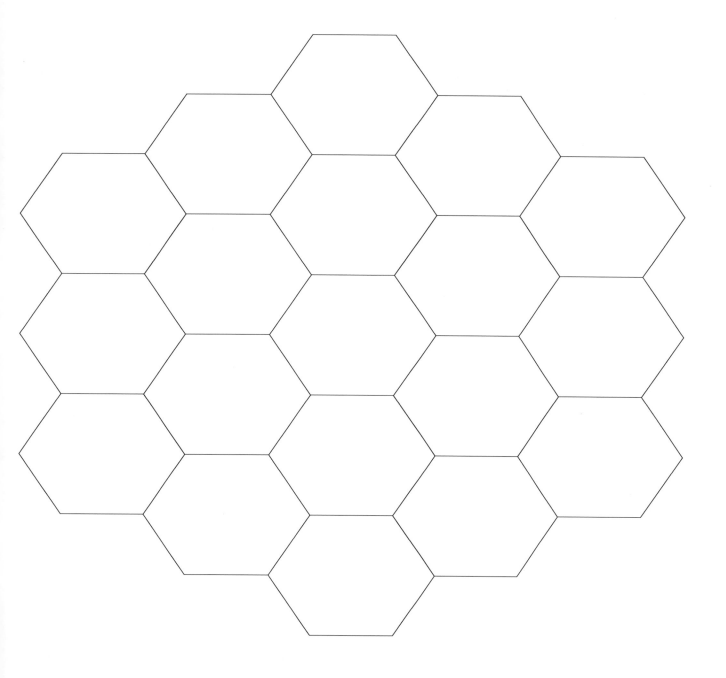

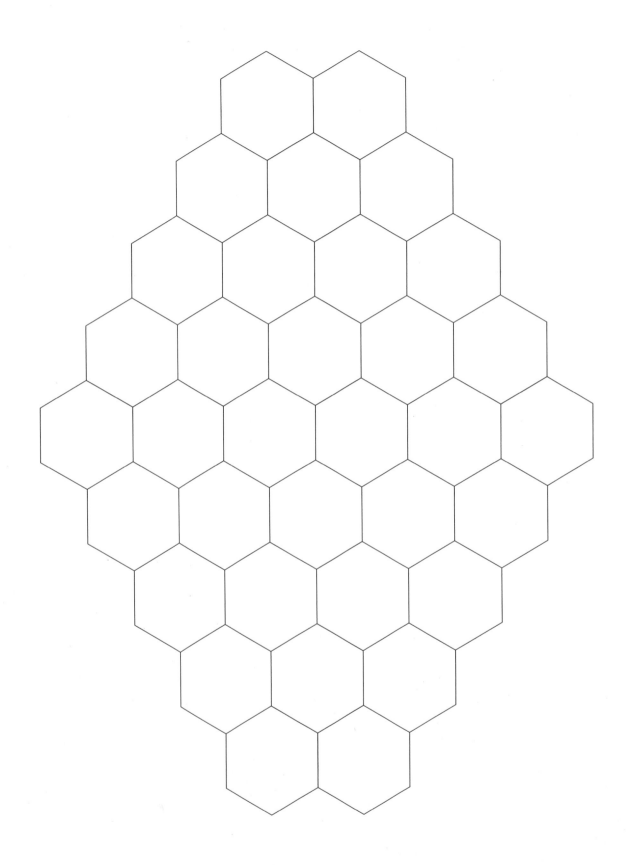